Bite Size Pieces

1st Bite

Create a Foundation
2nd Edition

Rebekah Hennes, R.D.

Real World Nutrition
5855 Green Valley Circle, Suite 210
Culver City, CA 90230
(562)895-0682   (310)861-5053 fax

Published and distributed in the United States by:
Lulu.com

The author does not dispense any medical advice or prescribe the use of any technique without the advice of a physician, either directly or indirectly. The intent of the author is to offer you advice on how to learn about and document your food intake and physical hunger and fullness cues. In the event that you use any of the information in this book for yourself, which is your constitutional right, the author and the publisher assume no responsibility for your actions.

Library of Congress Cataloging-in-Publication Data
Hennes, Rebekah

Bite Size Pieces, First Bite Create a Foundation
2nd Edition/ Rebekah Hennes
ISBN # 0-557-03885-5
ISBN 13 # 978-0-557-03885-5
1. Health
2. Nutrition
                    ISBN # 0-557-03885-5
              ISBN 13 #   978-0-557-03885-5
            1st Printing January 2009

TABLE OF CONTENTS

INTRODUCTION

Congratulations on taking the first bite. "Bite Size Pieces" are different from other self-help books. Each "bite" has been designed to be interesting, palatable, and digestible. Working through each book or "bite" will require some work as well, it is easy to read and accumulate knowledge but change really happens when you <u>DO</u> something different and you experience the difference for yourself.

"An ounce of practice
is worth a ton of misapplied knowledge."
R.S. Hukkerikar

Give yourself the opportunity to do the work. It will take time and you will need to be patient.

The assignments, information and experiences that are within these pages will provide you with the stability that is necessary for you to begin living a life free from diet and food obsessions.

The principles in this book are simple and yet they are not easy. You will be challenged to change the way you view food. Some beliefs that you have about food, which in the diet world may be touted as "fact", may prove to be questionable or even false when applied to a human being such as yourself. This work requires an open mind and the ability to take a risk.

HOW TO USE THE BOOK

Work through each "bite" thoroughly until you are able to digest the concepts and put the exercises and experiences into practice.

From past dieting, you may relate to food and your body negatively. By learning and exploring each "bite" you will improve your relationship with both food and your body.

It is a choice to let go of behaviors and habits that you no longer find helpful. Allow yourself to find excitement in the process of recovery. Learning who you are without your eating issues is priceless.

"The baby rises to its feet, takes a step,
is overcome with triumph and joy and falls flat
on its face.
It is a pattern for all that is to come!
But learn from the bewildered baby.
Lurch to your feet again.
You'll make the sofa in the end."

~Pam Brown

1st Bite ~

Create a Foundation

Humans build many things and we start building from the ground up. It will be the same in your recovery. The first "bite" or step is to create a foundation for your self. This begins by setting up a consistent schedule. Sticking to a schedule produces healthy ebbs and flows of dopamine and serotonin in your brain. Without enough of these "feel-good" chemicals you may find yourself with poor impulse control when it comes to food and possibly in other areas of your life as well.

Developing a base not only encompasses scheduling life activities, it also entails sitting down to eat regularly and supplementing your diet with nutrients so that any deficiencies, caused by years of dieting, are corrected. As part of creating a foundation you will also need to begin practicing imagery to increase your positive eating experiences. Lastly, you will need to start using alternative coping behaviors so that you do not turn back to dieting, purging, or another behavior in order to cope with stress and emotions.

The base that you develop will provide you with the foundation that is necessary for you to turn away from unnatural food behaviors and take steps towards living a life free from dieting and weight control. Possibly your past attempts to change behaviors were thwarted because you did not have a foundation to build upon. This may very well be the reason that you are still stuck using behaviors and continuing in habits that no longer serve your best interest.

1

SCHEDULE YOUR LIFE

Keeping to a consistent schedule is important on many fronts; the one we will discuss here has to do with two neurotransmitters (chemicals in your central nervous system), serotonin and dopamine. Serotonin plays an important role in the modulation of anger, aggression, body temperature, mood, sleep, sexuality, appetite, and metabolism, as well as stimulating vomiting. The chemical dopamine, because it elevates mood, increases motivation and enhances attention. Your central nervous system's use and dependence on dopamine and serotonin is natural and began when you were born. When you cried as a baby you received love and attention. When you received this love, your brain experienced a little boost of dopamine and serotonin. All animal brains are set up to want and need this "feel-good" chemical at regular intervals throughout the day...every day.

As we grow up, we begin to follow the rules and guidelines of the culture we are part of. Our western culture requires many of us to work long hours with few, if any, breaks. This lifestyle, in turn, does not provide the required dopamine and serotonin that the brain needs.

Therefore, on the way home from work, if you have a brain deficient in dopamine and serotonin, you may have poor impulse control which can lead you to crave convenient foods that are high in carbohydrate and fat. Stopping for donuts or a "fast food fix" will give your brain the precursors to dopamine and serotonin, but the calories and fat that they also provide may not be what your body biologically needs. This does not mean that you are addicted to sugar or fat. It does however, mean that your brain needs dopamine and serotonin just like the body and brain needs air and it will try to get it however it can.

Despite the culture you are part of, you will need to change some of the habits that you have acquired in order to experience a healthy relationship with food. You must provide yourself with a regular and consistent schedule so that your brain has serotonin and dopamine throughout the day. With normal levels of brain chemicals your impulses and cravings will not be driven by your western lifestyle and you will be able to listen to your body and feed it adequately. How do you know if you have a deficiency?

- You may have a dopamine deficiency if you are depressed and / or you crave high fat or sugary foods.

- You may have a serotonin deficiency if you have insomnia, depression, food cravings (specifically high sugar), you cannot control your behavior, or you experience aggressive behavior.

Additional Information:

Dopamine and serotonin are absent or deficient when we experience chronic stress. In our culture, chronic stress can also play a role in unnatural food cravings and as a result weight gain. Chronic stress causes an increase in insulin and cortisol levels which can cause weight gain in the mid-section, inhibit fat release, and lower serotonin levels. This, in turn, causes less impulse control, decreased dopamine levels, and increased food cravings.

The Eastern Spin:
Chi is the "vital life force." Three ways that you can cultivate chi are from food, exercise and doing activities that inspire you. The first way is from food. This is the most short-lived and fleeting form of chi. The second way to accumulate chi is through mind-body exercise such as tai-chi, NIA and yoga. The last and most long-lasting way to accumulate chi is by doing

things that inspire you, things that you are passionate about. On the following page, notice that sources of dopamine and serotonin are also activities that cultivate chi.

It may be that many people in our western culture are not doing enough activities that they are inspired by or passionate about, and to compound the problem they are not doing enough mind-body exercise. It makes sense that a person in this situation would crave and eat more food than their body actually needs to maintain a healthy weight. They are accumulating chi or life-force from the most inefficient source and are becoming over a natural weight as a result.

Increase your serotonin levels by:

- Using your senses
- Getting 1 hour of sunlight a day
- Getting enough sleep
- Keeping a positive attitude
- Being playful
- Exercising
- Keeping a consistent schedule
- Bonding with other people
- Eating omega 3 fatty acids
- Decreasing your daily stress

Increase your dopamine levels by:

- Doing things that you enjoy
- Excitement

What are you passionate about?

_____

_____

_____

_____

_____

_____

_____

_____

_____

_____

_____

_____

_____

_____

_____

_____

_____

_____

_____

What inspires you?

_____

_____

_____

_____

_____

_____

_____

_____

_____

_____

_____

_____

TIP- If you are so tied up in your food-weight hobby that you are unsure of what to write, try to remember what you used-to love to do.

After reviewing the previous pages, what are some activities that you would like to put into your daily/weekly schedule?

_____

_____

_____

_____

_____

_____

_____

_____

_____

_____

_____

_____

_____

_____

_____

_____

_____

USING FOOD TO INCREASE DOPAMINE AND SEROTONIN

The foods that are dopamine and serotonin precursors are also foods that have fat and protein in them and may be ones that you avoided eating due to their calorie content. By not eating these foods, you set yourself up for bingeing on high fat and high carbohydrate foods due to poor impulse control which was initiated by a dopamine and serotonin deficiency.

FOODS THAT INCREASE SEROTONIN

Tryptophan: Serotonin Precursor

Foods with 50-100 mg of tryptophan per exchange serving	Foods with >100 mg of tryptophan per exchange serving
Cheese: (blue, brick, Brie, Camembert, cheddar, Monterey, Muenster)	Almonds
Egg	Beef
Eggnog	Brazil Nuts
Fortified oat flakes	Canadian Bacon
Grapenuts and Grapenut Flakes	Cashew Nuts
Ham	Cheese: (Colby, Gruyere, Swiss, Cottage)

Foods with 50-100 mg of tryptophan per exchange serving	Foods with >100 mg of tryptophan per exchange serving
Life cereal	Chicken: (light and dark meat)
Mixed nuts	Goose
Peanuts	Hot Cocoa
Pecans	Lima Beans
Pine nuts	Milk
Pistachio nuts	Pheasant
Pork	Sesame Seeds
Sunflower seeds	Turkey Breast
Safflower seeds	
Soybeans	
Walnuts	
Wheat germ	
Yogurt	

Lists compiled from information obtained from a handout made by Monika Woolsey, M.S., R.D.

FOODS THAT INCREASE DOPAMINE

Phenylalanine: Nor epinephrine/ Dopamine Precursor

100-199 mg phenylalanine per exchange serving	200-299 mg phenylalanine per exchange serving	>300 mg phenylalanine per exchange serving
Bacon	Brazil nuts	Almonds
Cheese: parmesan	Cashew nuts	Cheese and yogurt
Corn	Hickory nuts	Chicken, light/dark
Hazelnuts	Kidney beans	Duck
Ham	Lima beans	Eggs
Lentils	Pine nuts	Goose
Navy Beans	Pork	Milk
Spinach	Safflower seeds	Peanuts
	Sesame seeds	Pistachio nuts
		Pumpkin seeds
		Soybeans
		Sunflower seeds
		Turkey
		Walnut

After reviewing the tables on the previous pages, are there any foods that you would like to start eating?

_____

_____

_____

_____

_____

_____

_____

_____

_____

_____

_____

_____

_____

_____

_____

_____

_____

REMEMBER- One way to have healthy levels of dopamine and serotonin is to stick to a schedule. You need to take yourself out on a walk as much as you need to take a dog out on a walk (at the same time everyday). You also, need to have a regular bed-time and wake-up-time as much as a child needs one.

If possible, at the same time...

1   Wake up.

2   Go to sleep.

3   Do exercise that you love.

4   Eat three meals and one to three snacks.

5   Call at least one person and talk about an important subject or something that you are passionate about.

6   Go outside and be in the sun for one-half to one hour a day.

7   Journal or meditate.

8   Take time to play, laugh, relax and breathe.

9   Be aware of your senses.

10 Read a positive quote or affirmation.

SCHEDULE YOUR LIFE (example)

TIME	ACTIVITY
8:00	Wake up
9:00	Breakfast *
9:30	Work
12:00	Lunch outside
12:30	Work
3:00	Snack and call a friend
3:30	Walk outside
4:30	Study/hobby
6:00	Dinner with friends or family
7:00	Journal
10:00	Sleep

*This example includes scheduled meals. There is more information about food and meal planning in the next chapter.

SCHEDULE YOUR LIFE

TIME	ACTIVITY

SCHEDULE YOUR LIFE

TIME	ACTIVITY

SCHEDULE YOUR LIFE

TIME	ACTIVITY
TIME	ACTIVITY

SCHEDULE YOUR LIFE

TIME	ACTIVITY

SCHEDULE YOUR LIFE

TIME	ACTIVITY

SCHEDULE YOUR LIFE

TIME	ACTIVITY

2

SCHEDULE YOUR MEALS

It is important that you SIT DOWN and eat at least three times a day. This will help you break the habit of chaotic eating. With regular meals your body will begin to experience that you are taking care of it. That, in itself, is an important step towards creating a healthy relationship with food.

It is also important to eat snacks. Your body requires food every three to four hours. If after four hours, you are still not hungry then you may have gotten uncomfortably full at the previous meal, you are ignoring a hunger signal, or you have forgotten what it feels like to be hungry.

SIT DOWN AND JUST EAT

It is important to sit down when you eat. Also, if you recognize "paired eating" (doing another activity while you are eating such as watching the television or working on the computer) it will be necessary to break that habit so that you can be fully aware of your eating experience.

REGULAR MEALS

It is very important to make regular meals a habit. After a few weeks your body will get used to you feeding it through-out the day so you will experience natural hunger and fullness signals instead of intense or non-existent signals. When you learn to recognize and honor your hunger and fullness signals you will not have to follow a schedule based on time. But right now, stick to a schedule and make it a habit.

TIP- Eating regularly will not make you fat. You can think of your body like a car. If you fuel it only one time a day then your gas tank will have to be bigger than if you fill-up three to five times a day.

Why should you eat every 3-4 hours?

1. Your metabolism will increase.

2. The more often you eat, the less hungry you will be. The less hungry that you are, the less you will want to over-eat. If you don't over-eat then you won't have the urges to restrict, binge, or purge.

3. You will be less likely to "over-snack." You may be surprised how much a person can consume by constantly snacking and never getting full-enough. If you are an "over-snacker" then you may actually consume less food if you sit down to eat three meals a day.

What if you are hungry every hour or two?

If you are hungry every one to two hours then you may be restricting calories, fat, protein, or carbohydrates or you may be emotionally eating.

What should I eat?

Right now don't stress about what you eat. Just make regular eating times a habit. If you are restricting fat, protein or carbohydrates then you might feel hungry more often and have cravings for the food that you restrict. You will have those cravings until your body receives all of the nutrients that it needs. It is a good idea to eat fat, protein and carbohydrate at all meals and most snacks.

Some food for thought...

Calories-

You need to eat enough calories so that you don't hurt your body. It takes 800 calories just to fuel your organs. If you don't eat at least 800 calories then you are compromising your heart, brain, etc. After the first 800 calories, your body needs 400 calories just for the "luxuries" of temperature, hair and nail growth, and skin health. On top of that, your body requires 300 calories for growth or pregnancy. Then your body needs 300 calories for energy which is used for mobility. "Mobility" includes getting out of bed and just doing normal life activities. If you exercise, your body needs even more. For example, when you eat 1200 calories and exercise for one hour, the amount of calories available for your body dips to less than 800. When this happens your body conserves energy, turns off your "luxuries," decreases your metabolism, and your organs begin to be affected.

Water-

An increase in weight does NOT always equal an increase in body mass. For example, have you ever restricted your food intake, consequently overeaten and then weighed yourself the next day? You may have noticed that your weight went up more than a couple pounds. If you consider that it takes 3500 calories in addition to 1500-2000 calories a day (for daily needs) to gain one pound, why is it that your weight went up more than a pound when you didn't eat 5000 calories?

Water is the reason. You didn't actually gain any body mass. You are 80% water by weight. Your cells need it and your entire body needs it.

If you are not eating enough carbohydrates (glucose), you aren't storing glucose for energy and your body begins to become dehydrated. Glucose is stored with water which is heavy, four water molecules to every glucose molecule. While you are maintaining a dehydrated and glucose deficient body you believe that you have lost weight when in fact you just don't have any stored glucose for energy, or water for hydration. Eventually, when you eat and drink what your body needs for regular glucose stores and hydration, you will gain water weight. It is not solid body mass, you never really lost any body substance in the first place so the weight increase is nothing to worry about.

The main point to digest is that scales are an inaccurate measure of real weight or health.

Protein-

Eat enough protein to keep your metabolism going. Protein is a building block. If you don't eat protein your body has no choice but to cannibalize your own muscle in order to make hormones, cells, etc. When your muscle mass decreases your metabolism goes down and it takes less calories to maintain your weight. It isn't a good cycle. It is important that you eat enough protein so that your metabolism can stay where it needs to.

Fat-

Eat foods with fat in them to feel satisfied longer. Body fat does not come from eating fat. Eating fat helps your body make cells, nerve coverings, etc. Fat can also help you feel full longer which means that you won't have so many cravings and you won't be hungry all the time.

When the fat-free craze was the "diet of choice" Americans ate more calories. Just because they were eating fat-free calories they were not protected from gaining weight. Eating fat free foods can actually make a person hungrier and cause them to eat more total calories than if they just ate the "real thing."

Another interesting fact is that when you sleep you burn a higher percentage of fat then when you exercise.

Carbohydrates-

Eat carbohydrates so that you have energy. Looking at carbohydrates from a "calorie perspective", it doesn't matter if the carbohydrate comes from broccoli or table sugar. Carbohydrates come from plants. Remember, sugar comes from sugar cane which is a plant. Your body breaks down carbohydrates into three molecules: glucose, galactose and fructose. Carbohydrates are in everything except fat and oil. Meat contains only a small amount.

Carbohydrates are nothing to be afraid of. They provide fuel for your body and your brain. They are energy made from plants, sun, water and carbon dioxide.

Just because a food tastes sweet in

your mouth, it does NOT mean that it's a bad food. Something that tastes sweet does not always have more carbohydrates in it than something that does not taste sweet. Sweet potatoes have the same amount of carbohydrates as regular potatoes

In the past you may have eaten a carbohydrate rich food in a way that felt out of control. This may have happened because you were not eating enough fat or protein. Also, it may also have happened because you had low levels of dopamine or serotonin, as explained in the previous chapter.

Osteoporosis-

Eat foods which contain calcium to keep your bones strong. Calcium makes strong bones but you have to be menstruating in order for your body to build bone mass. Osteoporosis is when your bones start to get holes in them where calcium should be. Some women's bones become so fragile that the bones in their hands break with a handshake.

You can be at risk for osteoporosis if you drink a lot of caffeine (coffee/soda), you have a small bone structure, you smoke, you don't menstruate and/or you don't eat dairy products. Even women in their twenties can develop osteoporosis and osteopenia. You only have until you are thirty years old to fill up your bones with calcium. After thirty you really can't increase your bone mass so therefore, you are not too young to care about your bones.

To help assure that you have strong bones you can drink or eat dairy foods or take calcium pills. But in order for you to not develop osteoporosis your weight has to be high enough for you to menstruate without the aide of birth control pills.

Laxatives-

Laxatives are a "no-no." First of all, they only get rid of (at most) 10% of what you eat. Secondly, when you eventually drink or eat, you may feel bloated because you are storing more body water. If you don't know any better, you may even think that you have gained weight or "feel fat." We'll talk about the "fat feeling" later.

Also, some people who abuse laxatives have to have their colons removed. After that, they go to the bathroom into a bag connected to a hole in the side of their body.

Metabolism-

Currently your metabolism may be as low as 70% of normal. When you are consistently eating enough, your metabolism will go back to a normal level. This process usually takes three weeks to six months. You will need to consistently eat more than 1200 calories everyday in order for your metabolism to return to normal. Remember, you didn't get to where you are today, overnight. It will take your body time, loving care, and a healthy relationship with food to get better. Be patient but take daily steps in the right direction and in time your metabolism will return to normal.

Weight-

By dieting, restricting, bingeing, and purging you are keeping your body at an unnatural weight. Your weight may be higher or lower than is natural for your body. In order to fully recover you CANNOT control your weight. You can never diet again.

If your body is maintaining a weight that is higher than before your eating disorder, your body will readjust to a healthy weight as you begin to practice the principles of natural or intuitive eating (which is not now). This is a process and you are on the first step. Give yourself at least two years of consistent practice for your body weight to reestablish itself.

Reasons that you could be at a higher weight than is natural for your body are: you restrict or avoid certain types of food, you feel deprived, you binge, you diet, you use laxatives, you emotionally eat, or you purge.

If you are underweight, you are probably always hungry but you don't let yourself eat. You can tell that you are hungry if you think about food a lot, read cookbooks or plan what you are going to eat days or weeks in advance. As you allow yourself to eat more, your body weight will readjust to a natural place. Depending on your metabolism and how wounded it is from dieting, it may take a lot of food (or a little) for you to gain weight.

When you give yourself full permission to eat, you may feel like you are bingeing and you may actually binge. It is biologically natural for that to happen.

When your weight is at a natural place your hunger and fullness levels will be natural as well and the urge to overeat will go away.

Exercise-

Exercising too much can be worse than exercising too little. Greater than one hour a day may be considered compulsive exercise. It is an impossible feat for your body to keep you healthy and repair cell damage (from exercise) if you are exercising too much. If you are eating less than what your body needs to support your exercise, and keep up with body maintenance activities, then your metabolism will go down, not up.

Diet Pills-

Stimulants are a bad idea. Any stimulant containing caffeine, Ma Huang or ephedra is very stressful to your nervous system and your heart. Not to mention that you can't take the pills for the rest of your life so you are likely to be extra hungry when you are no longer taking them. Some people have died as a result of taking products with stimulants. Side effects include high blood pressure, insomnia, arrhythmia, nervousness, tremors, headache, seizure, heart attack and kidney stones. Caffeine containing herbs (Guarana, Green Tea, and Kola Nut) and ephedra can alter the action of Prozac, Zoloft and Paxil.

COMMON QUESTIONS

What if I do not eat enough at a meal?

If you restrict your food intake you will be hungry again before three hours. If you don't eat enough food then your body will require that you eat more than five to six times a day.

What if I get too full at a meal?

If you get too full remember that it is okay. Your body will process what you ate and will tell you when and what to eat next. You will not become fat. Just wait until you are hungry again to eat. You may not recognize being hungry for several hours and that is okay.

Your body is like a parking meter. Feeding your meter less food will result in you having less time being satisfied. Feeding your meter more food will result in more time that you will have between meals.

WHAT TO EAT

Here are some ideas for meals. All of the following suggestions are conservative in calories. You may have gotten used to looking at a plate that is either filled with too little or too much (binge) food and now you need to change your idea of what a regular meal looks like. Please don't assume that you can only eat the foods on this list. These are just suggestions.

There are numbers on the following pages. Try not to get into the "number game" which includes calories, "right" and "wrong" serving sizes, etc. If you do become entrenched in the "number game," you will have to unlearn ALL of it in order to fully recover.

GENERAL INFORMATION

- 3 meals/day.

- 4 cups or more per day of colored veggies-mostly green,

- 20 grams of protein with each meal (3 eggs or 3 -4 oz meat or fish)

- Eat 1/3 of your food at breakfast

- Fruit-2/day

- Other carbohydrates- beans, rice, corn, etc.

- Dairy 2-4 servings/day

- Eat enough oil and fat. If you used to restrict fat and you feel like you are eating a ton of fat then it is enough.

- 40-50% carbohydrate, 30% fat (up to 66 grams of fat), and 20-30% protein (100-150 grams).

- Focus on getting enough protein and fat in a day (without restricting carbohydrates or calories) and you may notice a significant decrease in cravings.

BASIC DAY EXAMPLE

Breakfast: 3 eggs scrambled with veggies, avocado and black beans and a corn tortilla or two.

Lunch: a very large raw vegetable salad (regular dressing on it) with 4 ounces of meat or fish and a bowl of split pea soup or a side of roasted potatoes.

Dinner: 2 to 4 cups of sautéed veggies with tofu or feta cheese and 1 cup wild rice and brown rice pilaf.

Plus snacks of trail mix or fruit with cheese or nuts.

The example above can be found in: "DIET CURE"- Julia Ross, M.A.

SANDWICHES

Usually eating a full sandwich is appropriate (without changing its contents).
Or if you are at home:

Deli Sandwich
> 4 oz of deli turkey or chicken
> 2 slices whole wheat bread
> ¼ avocado or 1 slice of cheese or 2 slices of soy cheese or 1 tablespoon of mayonnaise
> lettuce and tomato
>> with either 1 ounce of Lay's sour cream and onion or BBQ chips or 1 Pepperidge Farm chocolate chip or oatmeal cookie

Peanut Butter Sandwich
> 2 tablespoons of peanut butter
> 1 tablespoon jam
> 2 slices bread
> 8 ounces (1cup) of soy milk or milk

BURGERS AND WRAPS

Veggie Burger
>   1 veggie burger
>   (or turkey burger-3 ounces)
>   1 bun
>   ¼ avocado or 1 slice cheese or 1
>   tablespoon mayonnaise or 2 slices soy
>   cheese
>   16 wheat thins or 1 cup of carrots with 2
>   tablespoons ranch
>   or 8 ounces of milk /soy milk
>   lettuce and sliced tomato

Cranberry Cream Cheese Wrap
>   12" flour tortilla
>   4 ounces of turkey or 8 slices of tofurkey
>   2 tablespoons cream cheese add the zest
>   of the orange (finely grate a little of the
>   outside of an orange)
>   1 tablespoon cranberry relish
>   grilled onion and spinach
>   fruit
>   (you can eat the orange that you used for
>   "zest" if you want to)

The meals above are examples from The Eating
Disorder Center of California.

PASTA

Pasta with red sauce and meat –
       at a restaurant or at home
       9" plate or a paper plate, full to the edges-
       ½ inch deep) or you can look at your
       hands, turn them into fists, and eat that
       amount which is about 2 cups.

OTHER

Traditional American Dinner
       Meat-
       a little bigger than the palm of your hand

       Vegetables- a fourth of the plate

       Carbohydrate (potatoes, rice, pasta)-
       ½ of your plate, ½ inch deep or the size
       of one balled-up-fist, which is 1 cup.

       Fat-
       The vegetables, meat and carbohydrate
       should be made with some fat (olive oil,
       butter, etc.)

SALAD

Spinach Salad
      2 cups of baby spinach
      4 oz of chicken or tofu
      2 tablespoons of balsamic dressing
      ½ of a pear
      1 tablespoon dried cranberries
      1 tablespoon toasted pine nuts
      Wheat roll (1 ounce)
      1 teaspoon butter

BBQ Ranch Salad
      2 cups romaine lettuce
      4 ounces chicken or tofu
      with KC Masterpiece BBQ Sauce to color
      and cover the meat/tofu
      ¼ cup tomatoes
      ¼ cup cucumber
      ¼ cup corn
      2 tablespoons ranch dressing
      1 Pepperidge Farm chocolate chip or
      oatmeal cookie

The meals above are examples from The Eating Disorder Center of California.

BREAKFAST

McDonalds
Sausage Mcmuffin with Egg or
Bacon Egg and Cheese Biscuit

At Home
    2 eggs or
    2 tablespoons peanut butter
    2 pieces toast
    1 tablespoon jam or ½ banana
    1 cup milk or
    container of light yogurt

Coffee House Scone

QUICK MEAL IDEAS

You really can eat fast food without gaining weight. The only reason that the "super-size-me" guy gained weight is because he over-ate a lot. The suggestions below are conservative, appropriate and convenient.

BAJA FRESH

3/4 of a chicken or steak burrito
Chicken bare burrito

KENTUCKY FRIED CHICKEN

3-piece Crispy Strips and Baked Beans
Roast Chicken Breast and Corn on the Cob

McDONALD'S

Hamburger, Garden Salad Shaker, herb vinaigrette salad dressing and small one-percent milk
Chicken McGrill and Garden Salad Shaker
Cheeseburger, 1% milk, fruit
Cheeseburger and small ice cream cone
Cobb salad and small ice cream cone
Hamburger and small French fries

PIZZA HUT

Two slices of The Edge "The Works" pizza and a tossed salad, 1Tbs ranch dressing

QUICK MEAL IDEAS

SUBWAY

6" ham/turkey sandwich on a deli roll plus one oatmeal raisin cookie

TACO BELL

Chicken Gordita Supreme and Mexican Rice

KOOKOOROO

BBQ Chicken Sandwich
Hand-carved Turkey Sandwich
Char grilled Chicken Bowl (BBQ Vinaigrette)
Tostada Bowl (shell optional) (BBQ Vinaigrette)
Southwest Bowl (BBQ Vinaigrette)
Spicy Ginger Garlic Bowl (BBQ Vinaigrette)
Sliced Turkey Plate with 2 of the following: black beans, creamed spinach, kernel corn, garlic potatoes, or stuffing and 1 of the following: green beans or Italian veggies.

It is better to "just eat" without counting or measuring. You will need to unlearn all of the "number information" and measuring habits in order to fully recover.

QUICK MEAL IDEAS

WENDYS
One item from this side…and one item from this side.

Jr Bacon	Yogurt /Granola
Jr Cheese Burger Deluxe	Caesar Salad
Large Chili	Small Frosty
Fries	Kids meal fries
Sour Cream Potato	
Ultimate Chicken Grill	
Cranberry Pecan Salad	

or

Broccoli and Cheese Potato
Fish Sandwich
10 piece Chicken Nuggets
Mandarin Chicken Salad (with the dressing it comes with).
Roasted Turkey Basil Pesto Sandwich

A note on fast food:
If you are consistently eating fast food and your body doesn't feel very good, then it may be a good time to eat some "home cooked meals."

SNACK IDEAS

2/3 cup yogurt and fruit

3/4 cup yogurt and 1/4 cup granola

2 string cheese and fruit

2 string cheese and 5 triscuits or 8
wheat thins

2 snickers bars (fun size)

yogurt honey peanut balance bar and
fruit or 4 oz fruit juice

3/4 cup chocolate or rice pudding

2 Reese's peanut butter cups

Apple or 1/2 banana and 2 tbsp
peanut butter

1/4 cup trail mix

3/4 cup dreyers ice cream

2 full sheets regular graham crackers
and 1 cup milk or soy milk

Luna bar and fruit or 4 oz juice

10 mini pretzels and 2 string cheese

2/3 cottage cheese and fruit

1 oz potato chips

1/4 cup any kind of m&ms
or (one package- not king size or
snack size)

16 wheat thins and 4 tbsp hummus

2/3 cup hummus and 1 c carrots

SCHEDULE YOUR MEALS

Write down when **and what** you will eat

TIME	MEAL/SNACK

SCHEDULE YOUR MEALS

Write down when **and what** you will eat

TIME	MEAL/SNACK

SCHEDULE YOUR MEALS

Write down when **and what** you will eat

TIME	MEAL/SNACK

SCHEDULE YOUR MEALS

Write down when **and what** you will eat

TIME	MEAL/SNACK

SCHEDULE YOUR MEALS

Write down when **and what** you will eat

TIME	MEAL/SNACK

73

3

SUPPLEMENT FOR DEFICIENCIES

You may have some nutrient deficiencies due to dieting. This can be the case even if your body weight is higher than natural. Besides the basic recommendations on the next couple of pages, other supplements can be helpful…especially if you believe in them (the placebo effect is extremely powerful). Bare in mind, you do need to be cautious with supplements.

Not all supplements work and not all supplements contain what is on the label (supplements are not regulated). According to Dr Andrew Weil you can supplement your food intake with the following:

BASIC VITAMINS AND MINERALS

Supplement your diet with the following antioxidant cocktail:

- Vitamin C, 200 milligrams a day.

- Vitamin E, 400 IU of natural mixed tocopherols (d-alpha-tocopherol with other tocopherols, or, better, a minimum of 80 milligrams of natural mixed tocopherols and tocotrienols).

- Selenium, 200 micrograms of an organic (yeast-bound) form.

- Mixed carotenoids, 10,000-15,000 IU daily.

- Take daily multi-vitamin supplements that provide at least 400 micrograms of folic acid and at least 1,000 IU of vitamin D. Containing no iron and no preformed vitamin A (retinol).

- Take supplemental calcium, preferably as calcium citrate. Women should supplement with 500-700 mg daily, for a total daily intake of 1,000-1,200 mg from all sources.

OTHER BASIC SUPPLEMENTS

- If you are not eating oily fish at least twice a week, take supplemental fish oil, in capsule or liquid form, 1-2 grams a day. Look for molecularly distilled products certified to be free of heavy metals and other contaminants.

- If you are not regularly eating ginger and turmeric, consider taking these in supplemental form. Turmeric and ginger are good spices to add to rice.

- Add CoQ10 to your daily regimen: 60-100 milligrams of a soft gel form taken with your largest meal.

SUPPLEMENTS FOR GI PAIN

If you have abdominal pain, indigestion, or discomfort make sure that you are seeing a gastro-intestinal (GI) doctor and a dietitian. There are some gastro-intestinal issues that can arise from disordered eating and additional supplements or medicines may be warranted.

OTHER SUPPLEMENTS

Below are some of the basic supplements recommended by Julia Ross, MA. She has over twenty years of experience helping clients with disordered eating and recognizes the helpfulness of the following supplements.

Type	Amount	What for?
Zinc Tally (liquid zinc)	40 mg 3x day early morning, mid-morning and mid-afternoon	1/3 of bottle with each meal until it starts tasting bad. Once a month test yourself to make sure that it still tastes awful.
Complete Digestive Enzymes with HCL	1 or 2 with breakfast, lunch and mid-afternoon snack.	If any food makes you want to throw up discuss this with your doctor. You may need a prescription to help with stomach motility.
Vitamin C	Emergen-C Lite Packets 4x day take early morning and with three snacks.	
B Complex	3 x day, early morning, mid-morning and mid-afternoon	

4

USE IMAGERY

"Imagery (visualization) can relieve pain, speed healing and help the body subdue hundreds of ailments including depression, anxiety and eating disorders. The power of the mind to influence the body is quite remarkable. Although it isn't always curative, imagery can be helpful.

Imagery is also at the center of relaxation techniques designed to release brain chemicals that act as your body's natural brain tranquilizers which lower blood pressure, heart rate and anxiety levels. Because imagery relaxes the body, doctors specializing in imagery often recommend it for conditions such as headaches, chronic pain in the neck and back, high blood pressure, and spastic colon.

Studies have shown that imagery can improve the immune system, quality of life, decrease pre-menstrual symptoms, decrease mood swings, ease pain, lift depression, lower blood pressure, slow heart rate and help treat insomnia, obesity and phobias."

More Information on Imagery is available online:

www.holisticonline.com/guided-imagery.htm

IMAGERY AND EATING

I have often requested that clients practice imagery in order to change behavior. If a client's goal is to eat pizza at school and when she tries, she throws it away before taking a bite, I will ask her to replay the scenario many times in her mind in order to see herself eating the pizza. If she cannot see herself eating the pizza in her imagination, it is unlikely that she will be able to eat it in real-life.

Besides, if it takes thirty times to actually change a behavior then you might as well replay the scenario thirty times in your head without even taking a bite. Otherwise, you have to actually eat the food thirty times in order to get the same result. It is easier, less stressful, saves money and decreases the number of anxiety producing experiences that you will have to through, if you just practice imagery. Besides, (speaking the language of the eating disorder) you are eating twenty-nine less real pieces of pizza if you do it in your head first.

My favorite example of imagery in action happened by accident. Mandy taught herself imagery because she was bored and had a headache. Mandy was severely anorexic and had lost so much weight that her intestines collapsed with the pressure of her aorta. An illness called Superior Mesentery Artery syndrome (SMA). Mandy was hospitalized and was unable to eat for six weeks, instead she was tube fed. Every night, the tube coming out of her nose would be hooked up to a feeding machine and she would sit in a rocking chair for hours being re-fed. The feeding process would give her a headache so she would keep her eyes closed. SMA syndrome was a wake-up call, she was determined to learn how to live without anorexia, to eat and have a good life.

During Mandy's headaches she began to vividly visualize herself at her favorite Mexican restaurant with her family, ordering and eating a burrito. During her first attempt at imagery, she could not get through the door of the restaurant. With time, she was able to order the burrito. In later visualizations, she was able to eat and enjoy the burrito without anxiety, loving the company of her family and leaving the restaurant happily.

After six weeks practicing imagery, she had the naso-gastric tube removed and she entered the day-patient program where I was the dietitian. On our first meal outing Mandy was not anxious, she was able to order a challenging item, she was a joy to be around and she reported no issues after the meal.

I was unaware that she had practiced imagery and I was sure that Mandy was hiding her feelings or doing eating disorder behaviors when she went home at night. I questioned her, "Why does it seem easy for you to eat?" She described the six weeks that she had used imagery and it all made sense. Mandy was a shining example of how a person so sick and so eating disordered could use her mind to change her reality. I am still amazed. I have followed her progress for years and she has truly recovered.

COVERT BEHAVIORAL REHEARSAL

The type of imagery practice that she used was called Covert Behavior Rehearsal (CBR). In CBR imagery, the individual systematically visualizes the desired coping behavior. This technique has seen much use in sports.

One study was done with basketball players. One group practiced shooting hoops and the other group imagined themselves shooting hoops and making the shot. The group that imagined that they were making the baskets improved more than the group who physically practiced the behavior.

PRACTICE CBR

Try to practice imagery two or three times a day. Most people find it easiest to do in bed, in the morning and at night before falling asleep, though with practice you'll be able to visualize whenever and wherever the need arises.

Write down a challenging place to eat.

_____

_____

_____

_____

_____

List some challenging foods to eat.

_____

_____

_____

_____

_____

_____

_____

_____

_____

_____

_____

Choose a visualization that you would like to practice using imagery. Write a detailed description of the outcome that you will be visualizing.

_____

_____

_____

_____

_____

_____

_____

_____

_____

_____

_____

_____

_____

_____

_____

_____

_____

Give yourself a couple of weeks practicing imagery daily. What was your experience like?

_____

_____

_____

_____

_____

_____

_____

_____

_____

_____

_____

_____

_____

_____

_____

_____

_____

_____

5

USE OTHER BEHAVIORS

Do NOT make yourself purge, restrict your food intake and try not to feel guilty. It isn't necessary and it will not help. Purging will cause insulin and blood sugar fluctuations which will make you hungry right away and can make you gain weight even without eating again. Guilt will only make the situation worse. Restricting is not necessary because if you listen to your body it will tell you when to eat.

Here are some reasons that you may use unhealthy eating or purging behaviors:

Reasons:
You are not feeling what you need to feel.
You are overwhelmed with life, work, or school.
You let yourself get too hungry.
It is a habit.
You feel out of control.
You are not saying what you need to say.
You get too full and you think you have to purge so that you don't gain weight.
Other:

It will be great when you are able to recognize exactly why you want to use compensatory behaviors and deal with the actual feelings in the moment instead of avoiding the feelings by acting out on your behaviors. Soon you will recognize what emotions you are not comfortable feeling and with time you will learn to feel them.

It is okay to feel sad, angry, or happy. There is not a bad emotion there are only different emotions. All emotions have energy that you can use and learn from. Just don't get caught up in the emotion or try to suppress it. Let yourself feel the emotion and then it will go away.

Until you are able to experience and deal with situations and emotions directly and in the moment, when you recognize yourself wanting to use a disordered behavior take fifteen minutes to try to make yourself feel better by doing other activities.

Pick three activities from the following list that you will try for five minutes each. After fifteen minutes of really trying to feel better you can choose to compensate if you still desire to. However, you really have to try to feel better. Don't just wait for the fifteen minutes to be over.

FIVE MINUTE ACTIVITIES

Place a check-mark in the column to the left of the item that you are going to try. Choose at least three.

	Read		Listen to music
	Breathe		Talk
	Feel what you need to feel.		Call a friend
	Say what you need to say.		Dance
	Close your eyes.		Do a puzzle
	Brush your pet.		Vacuum

Write.		Wash your car.	
Take a bath.		Plant something	
Sit outside.		Cry	
Look at the sky.		Give yourself a present.	
Pray		Reorganize anything	
Meditate		Stretch.	

Write about your ability to help yourself feel
better. Have there been times that you were able
to stop yourself from bingeing, restricting, or
purging?

_____

_____

_____

_____

_____

_____

_____

_____

_____

6

CONCLUSION

There is no right or wrong path to recovery. This is your journey and you will need to take it at a pace that is right for you. You are worthy of being happy and having a life free from your eating disorder.

"You will not grow if you sit in a

beautiful flower garden,

but you will grow if you are sick,

if you are in pain, if you experience losses,

and if you do not put your head in the sand,

but take pain and learn to accept it,

not as a curse or punishment

but as a gift to you

with a very specific purpose."

-Elizabeth Kubler-Ross

7

DAILY JOURNAL

At first, the only way for you to really know if your body is getting enough calories, as well as the specific types of food that it needs, is to begin using a food journal to gain insight into your food intake. After you have kept a record of your food intake for a week, you can ask for advice from a dietitian who will have the information that you have collected to be able to offer you suggestions, as appropriate.

This food journal has a column for you to

write down what and when you eat. There is also a column for you to document what emotion you are feeling prior to eating. Remember that your emotion is not dictated by what you eat. Write down instead, what you felt before eating. It may be a mood from the moment in general.

On the right side of this food journal, there are two columns for you to document eating disordered behavior. If you have an urge to binge, restrict, or purge you can note the urge in the second to the last column. In the last column, you can document if you follow through on any of those behaviors. Also, if you do binge (or over-eat) it is very helpful and freeing, albeit challenging, to write down the foods that you eat during the binge.

On the following page there is a table that lists emotions. This can be very helpful to look at when you are not sure what you are feeling.

EMOTION LIST

SAD	HAPPY	ANGRY	AFRAID
Abandoned	Amazed	Abused	Ambivalent
Agonized	Amused	Aggressive	Anxious
Apologetic	Calm	Alienated	Apprehensive
Burdened	Cherished	Angry	Bewildered
Desperate	Comfortable	Apathetic	Cautious
Disappointed	Confident	Appalled	Confused
Discouraged	Content	Blamed	Cowardly
Distant	Determined	Bitter	Disoriented
Disregarded	Delighted	Bored	Fearful
Embarrassed	Eager	Controlled	Frantic
Empty	Ecstatic	Disapproving	Frightened
Foolish	Exilarated	Disgusted	Hesitant
Forgotten	Free	Enraged	Insecure
Grief	Fulfilled	Envious	Panicked
Hopeless	Happy	Exasperated	Paranoid
Humiliated	Hopeful	Frustrated	Perplexed
Hurt	Important	Furious	Puzzled
Hysterical	Joyous	Guilty	Restless
Impotent	Loving	Hostile	Scared
Isolated	Loose	Horrified	Suspicious
Jinxed	Mellow	Impatient	Threatened
Lonely	Mischievious	Indifferent	Timid
Lost	Nurturing	Irritated	Torn
Miserable	Optimistic	Lethargic	Uncertain
Neglected	Peaceful	Manipulated	
Overlooked	Playful	Negative	
Regretful	Protective	Ornery	
Rejected	Proud	Resentful	
Upset	Relieved	Shocked	
Withdrawn	Respected	Smothered	
Worthless	Satisfied	Stubborn	
Vulnerable	Sympathetic	Victimized	

EXAMPLE

Date:				
Supplements Taken:				
Imagery Practiced:				
Other Coping Behaviors Used:				
Time	Emotions	Food	Urge to B/R/P	I Did B/R/P
6:00 AM	sad	1 c cereal, 1c milk	no	no
9:00 AM	disappointed	apple	no	no
12:00 PM	Guess it's time to eat. I can't study. Frustrated.	peanut butter sandwich	no	no
3:00 PM	Annoyed. I am thinking about food.	Apple	no	no
6:00 PM	Anxious	Pot of coffee, salad with vinegar	no	no
9:00 PM	Depressed	Diet soda	no	no

Date:				
Supplements Taken:				
Imagery Practiced:				
Other Coping Behaviors Used:				
Time	Emotions	Food	Urge to B/R/P	I Did B/R/P

107

Date:					
Supplements Taken:					
Imagery Practiced:					
Other Coping Behaviors Used:					
Time	Emotions	Food		Urge to B/R/P	I Did B/R/P

Date:					
Supplements Taken:					
Imagery Practiced:					
Other Coping Behaviors Used:					
Time	Emotions	Food		Urge to B/R/P	I Did B/R/P

109

Date:				
Supplements Taken:				
Imagery Practiced:				
Other Coping Behaviors Used:				
Time	Emotions	Food	Urge to B/R/P	I Did B/R/P

Date:					
Supplements Taken:					
Imagery Practiced:					
Other Coping Behaviors Used:					
Time	Emotions	Food		Urge to B/R/P	I Did B/R/P

111

Date:				
Supplements Taken:				
Imagery Practiced:				
Other Coping Behaviors Used:				
Time	Emotions	Food	Urge to B/R/P	I Did B/R/P

Date:					
Supplements Taken:					
Imagery Practiced:					
Other Coping Behaviors Used:					
Time	Emotions	Food		Urge to B/R/P	I Did B/R/P

Date:				
Supplements Taken:				
Imagery Practiced:				
Other Coping Behaviors Used:				
Time	Emotions	Food	Urge to B/R/P	I Did B/R/P

Date:					
Supplements Taken:					
Imagery Practiced:					
Other Coping Behaviors Used:					
Time	Emotions	Food		Urge to B/R/P	I Did B/R/P

Date:				
Supplements Taken:				
Imagery Practiced:				
Other Coping Behaviors Used:				
Time	Emotions	Food	Urge to B/R/P	I Did B/R/P

Date:					
Supplements Taken:					
Imagery Practiced:					
Other Coping Behaviors Used:					
Time	Emotions	Food		Urge to B/R/P	I Did B/R/P

117

Date:				
Supplements Taken:				
Imagery Practiced:				
Other Coping Behaviors Used:				
Time	Emotions	Food	Urge to B/R/P	I Did B/R/P

Date:					
Supplements Taken:					
Imagery Practiced:					
Other Coping Behaviors Used:					
Time	Emotions	Food		Urge to B/R/P	I Did B/R/P

Date:				
Supplements Taken:				
Imagery Practiced:				
Other Coping Behaviors Used:				
Time	Emotions	Food	Urge to B/R/P	I Did B/R/P

Date:				
Supplements Taken:				
Imagery Practiced:				
Other Coping Behaviors Used:				
Time	Emotions	Food	Urge to B/R/P	I Did B/R/P

Date:				
Supplements Taken:				
Imagery Practiced:				
Other Coping Behaviors Used:				
Time	Emotions	Food	Urge to B/R/P	I Did B/R/P

Date:				
Supplements Taken:				
Imagery Practiced:				
Other Coping Behaviors Used:				
Time	Emotions	Food	Urge to B/R/P	I Did B/R/P

Date:				
Supplements Taken:				
Imagery Practiced:				
Other Coping Behaviors Used:				
Time	Emotions	Food	Urge to B/R/P	I Did B/R/P

Date:				
Supplements Taken:				
Imagery Practiced:				
Other Coping Behaviors Used:				
Time	Emotions	Food	Urge to B/R/P	I Did B/R/P

Date:				
Supplements Taken:				
Imagery Practiced:				
Other Coping Behaviors Used:				
Time	Emotions	Food	Urge to B/R/P	I Did B/R/P

Date:				
Supplements Taken:				
Imagery Practiced:				
Other Coping Behaviors Used:				
Time	Emotions	Food	Urge to B/R/P	I Did B/R/P

Date:				
Supplements Taken:				
Imagery Practiced:				
Other Coping Behaviors Used:				
Time	Emotions	Food	Urge to B/R/P	I Did B/R/P

Date:				
Supplements Taken:				
Imagery Practiced:				
Other Coping Behaviors Used:				
Time	Emotions	Food	Urge to B/R/P	I Did B/R/P

FREQUENTLY ASKED QUESTIONS

Emotions

- How do I know if I am emotionally eating?

Emotional eating is eating when the body is not hungry, eating to decrease emotional sensations, or eating to distract your self from thoughts or feelings. It is filling your body with food instead of feeling your emotions. Use your food journal and monitor to document the emotions that you are experiencing prior to eating. You can use the flow chart (in the appendix) to help you differentiate physical and emotional hunger. If you do find that you are emotionally eating, the next step is for you to practice alternative coping methods. The "gold-standard" method is dealing with your emotions head-on such as experiencing the emotion, writing about it, or talking about it. The second method entails doing some nurturing activity such as taking a bath or listening to music. The third coping method is to distract your self from the situation by doing an activity such as reading a book or watching a movie. Coping with an emotion or situation without using food can be similar to that of breaking any addiction. It is not easy.

- When I get upset, I do not want to eat. Emotional restricting is similar to emotional eating. It has different means but the same end. If you are faced with a challenging situation, or emotion, and you do not eat or you eat very little as a way to cope with, or numb, your feelings then you are using restricting as a coping method. If this has become a habit for you, you will need to learn to separate your body cues from your emotions, and you will need to eat even if you are upset.

Fullness

- I am a little over-full after eating a special meal.

It is perfectly normal to get a little over-full on occasion. You will want to tune back into your hunger level as soon as you can without guilt, so that you do not feel the need to "finish the job" by eating more and /or purging.

- I just ate and I am uncomfortably full. What do I do?

Nothing, just wait. You will get hungry again, but it will take longer for this to happen. It is like you put a quarter into a parking meter instead of a dime. You just have more time to do other activities before you come back to feed the meter.

- I am consistently over-eating.

  What should I do?

Check in with yourself to see if you are emotionally eating, and learn other coping skills. If you consistently over-eat, then you will gain weight. But, if you get overfull less than twenty percent of the time your weight will probably not be affected. If your weight does go up because you get too full, too often, your weight will go back down after you have perfected your ability to stop eating at a satisfied level. You must be patient with the learning process.

Hunger

- I am hungry all the time.

If you are malnourished, underweight, or have been dieting you may experience a constant hunger sensation due to your body wanting to either gain weight, or to take in adequate nutrients such as protein, carbohydrates, or fat. It is not uncommon for you to feel like you want to overeat, or even experience overeating after dieting. This extreme hunger will decrease once you are back to your natural set-point weight, and are eating enough food so that your body does not react as if it is being starved.

- I have no idea if I am hungry or full.

You will usually begin to monitor your hunger and fullness levels inaccurately. You may think that you are very full when you eat a challenging food, or you may believe that you are full all of the time. It is important that you see a dietitian who can help you sort out your distorted translation of body cues. A dietitian will help you recognize what body sensations indicate that you are hungry and full. For example, if you think that you are at a "5", and you believe that you are not hungry prior to eating a meal, but you also are preoccupied with food, you have a headache, or you are feeling irritable, you are actually at a "2" on the hunger and fullness scale. On the other hand, if you believe that you are at an "8", but you are not uncomfortably full, then you are actually at a "6" on the hunger and fullness scale. As you can see, you will need some help understanding your hunger and fullness cues.

Medication
- Medication and Hunger

Some psychiatric medications trigger abnormal hunger and fullness signals. It is important for you to discuss your concerns with your psychiatrist, and to let your dietitian know what medications you are taking.

Purging

- Right after I purge I am very hungry. Why?

When you binge, insulin is dumped into the bloodstream. When you purge, some insulin stays in the veins to break down the glucose from the binge, and take it into the body for use. Because purging removes some of the food that the insulin was prepared to work on, the remaining insulin drops the blood sugar even lower which will cause you to feel hungry, and may trigger an episode of hypoglycemia. This is one of the causes of the continuous binge-purge-binge-purge cycle.

Types of food

- Fat, Hunger, and Satiety

Another factor contributing to your ability to stop eating when you are at a satisfied level has

to do with your intake of fat, protein, and carbohydrate. You must be eating an adequate amount of fat and protein in order to feel satisfied. If you do not provide your body with the food that it needs you will experience a constant hunger. Also, you may not even recognize that you are hungry, because you are not able to recognize a fullness sensation for comparison. Some people describe this feeling as being constantly agitated.

When you add a significant amount of fat to your diet (approximately 30% or greater), you will be able to recognize satiety. After that, you will be able to differentiate hunger from fullness.

- I am a vegetarian.

Does protein really matter?

Absolutely! Many people say that once they eat enough protein (3-4oz meat at a meal) they recognize being satisfied for a longer period of time. You will want to ask your dietitian if you are getting enough protein…especially if you are a vegetarian.

- I am scared to eat fat.

Does eating fat really matter?

You bet. You may think that you cannot control yourself around carbohydrates, but the truth may be that you don't eat enough fat. Contemplate this exaggerated example: if a slice of bread contains 1 gram of fat, your body needs 60 grams, and you refuse to eat foods that have a high amount of fat in them, your body will not let you rest until it triggers you to eat 60 slices of bread just for the fat (6,000 calories). On the other hand, if you ate 4 slices of bread with peanut butter (1/3 cup) you will eat far less calories because you will eat less than 60 slices of bread to ingest 60 grams of fat (approximately 1,000 calories total).

By restricting fat, you are causing your body to go on a hunt for inefficient sources of fat, and you will end up eating more calories in the process. Trying to replace fat with carbohydrates is like trying to give a car only gas when it is low on oil. Your best bet is to give yourself permission to eat all foods, and trust that your body will tell you what food it needs, when it needs it.

BIBLIOGRAPHY

Costin, C. *The Eating Disorder Sourcebook.* New York: McGraw-Hill, 2006.

Hansen, V, and S Goodman. *The Seven Secrets of Slim People.* Carlsbad: Hay House, Inc., 1999.

Hennes, R. *One Day at a Time. Food Journal and Hunger Fullness Monitor.* CCH Services, Inc., 2007.

Pawlak, L. *Stop Gaining Weight.* Kanab. Jeblar, Inc. 2005.

Reiff, DW, KKL Reiff. *Eating Disorders: Nutrition therapy in the recovery process._*Aspen Publishers, 1998.

Ross, J. *The Diet Cure: The 8-Step Program to Rebalance Your Body Chemistry and End Food Cravings, Weight Problems, and Mood Swings-Now,* Penguin Group USA, 2000.

Tribole, E, E, Resch. *Intuitive Eating: A Revolutionary Program that Works.* 2nd Ed. New York: St. Martin's Griffin, 2003.

Weil, A. *Eating Well for Optimum Health.* New York: Harper Collins, 2001.

BOOKS...
TO HELP YOU ON YOUR JOURNEY

- The Seven Secrets of Slim People
  Vikki Hansen, MSW and Shawn Goodman

- Intuitive Eating: A Revolutionary Program
  that Works
  Evelyn Tribole, MS, RD and Elyse Resch,
  MS, RD, FADA

Rebekah Hennes, R.D.

Rebekah Hennes has worked at the Eating Disorder Center of California, Pepperdine University, and Center for Change. She also worked at the Monte Nido Treatment Center. She is currently in private practice in Culver City, California. Rebekah recently edited and co-authored the book "Real World Recovery - Intuitive Food Program Curriculum for the Treatment of Eating Disorders" and authored "Return to Health" and the food journals "One Day at a Time" and "breathe." She was a contributor to the "Eating Disorder Sourcebook", 3rd Edition, and was a reviewer of the American Dietetic Association position paper on Eating Disorders. Rebekah pioneered the Intuitive Food Program at the Center for Change in 1999.

You can reach Rebekah at:
rebekah@realworldnutrition.org.
For more information, please visit the
Real World Nutrition website at:
www.realworldnutrition.org

www.ingramcontent.com/pod-product-compliance
Lightning Source LLC
Chambersburg PA
CBHW061314280526
45784CB00002B/985